Cartoon Figural Toys

Jameson Scott and Jim Rash

Schiffer Publishing Ltd®

4880 Lower Valley Road, Atglen, PA 19310 USA

I0809305

Title page:
Top right: Tigger™ vinyl figure-5.5"-Disney Park™ exclusive-1990-L-$6
Bottom right: Daffy Duck™ latex squeeze toy-6.5"-Lanco (Spain)-1988-BG-$10
Bottom left: Bump in the Night Mr. Bumpity™-8"-Green Grass Prod.-L-$8
Copyright page:
Mighty Mouse™ vinyl figure-6"-Dakin-1977-BG-$200

Library of Congress Catalog Card Number: 99-60442

Designed by Bonnie M. Henlsey
Layout by Randy L. Hensley
Type set in Americana XBd BT/Humanst 521 BT

ISBN: 0-7643-0832-7
Printed in China
1 2 3 4

Published by Schiffer Publishing Ltd.
4880 Lower Valley Road
Atglen, PA 19310
Phone: (610) 593-1777; Fax: (610) 593-2002
E-mail: Schifferbk@aol.com
Please visit our web site catalog at **www.schifferbooks.com**

In Europe, Schiffer books are distributed by Bushwood Books
6 Marksbury Avenue Kew Gardens
Surrey TW9 4JF England
Phone: 44 (0)181 392-8585; Fax: 44 (0)181 392-9876
E-mail: Bushwd@aol.com

This book may be purchased from the publisher.
Include $3.95 for shipping. Please try your bookstore first.
We are interested in hearing from authors with book ideas on related subjects.
You may write for a free printed catalog.

Disclaimer

This book is in no way intended to infringe on the intellectual property rights of any party. All products, brands, characters, and names represented are trademarks or registered trademarks of their respective companies, all rights reserved.

The information in this book is derived from the authors' independent research and is not authorized, furnished or approved by any of the companies mentioned in the book.

Below are all trademarked characters and titles exhibited in this book, listed by the companies which own them (copyrights may apply):

Bagdasarian Productions: Alvin™; Alvin and the Chipmunks™; Simon™; Theodore™; David Seville™.
Bay View Savings Bank: Bay View Eagle Bank™.
Bob Clampett: Beany and Cecil™; Dishonest John™.
Bob's Big Boy Restaurants/Marriot: Bob's Big Boy™.
Bristol Meyer: Speedy Alka Seltzer.™.
CBS TV: Captain Kangaroo™.
Creston Studios: Crusader Rabbit™; Rags the Tiger™.
Dakin: CartoonTheater™; Goofy Grams™; Dream Dolls™; Dream Pets™; Squeakies™.
Disney/The Walt Disney Company: Bambi™; Bashful™; Bongo™; Captain Hook™; Croc™; Daisy Duck™; Dewey™; Donald Duck™; Dopey™; Dumbo™; Fantasia™, Fun and Fancy Free™; Goofy™; Louie™; Ludwig Von Drake™; Mickey Mouse™; Minnie Mouse™; Mowgli™, Peter Pan™; Pinocchio™; Pluto™; Snow White and the Seven Dwarfs™; Thomas O'Malley™; Tinkerbell™; Tramp™; Trusty™.
Dr. Seuss Enterprises L.P.: Birthday Bird™; Busby Tasselated Afghan™; Cat in the Hat™; Dr. Seuss™; Gowdy the Dowdy Crackle™; Grickily Gractus™; Grinch™; Hedwig™; Horton the Elephant™; Merry Menagerie™; Norval the Bashful Blinket™; Roscoe the Many Footed Lion™; Tingo the Noodle-Topped Stroodle™; Yertle the Turtle.™.
Douglas Oil Company: Freddy Fast Gas™.
Felix the Cat Productions, Inc.: Felix the Cat™.
General Mills: Boo Berry™; Cinnamon Toast Crunch™; Count Chocula™; Frankenberry™; Fruit Brute™; Lucky Charms™; Sonny™; Cocoa Puffs™; Trix™; Twinkles™.
Golden Books Family Entertainment: King Leonardo™, Tooter the Turtle™; Underdog™.
Green Grass Productions: Bump in the Night™.
Hanna-Barbera: Atom Ant™; Augie Doggie™; Autocat™; Baby Puss™; Bamm Bamm™; Banana Splits™; Barney Rubble™; Benny the Ball™; Betty Rubble™; Bingo™; Blip™; Brain™; Boo Boo™; Boo the Bulldog™; Breezly™; Bubi Bear™; Castor Beaver™; Chattanooga Cat™; Chopper™; Cindy Bear™; Dick Dastardly™; Dino™; Dixie™; Droop-A-Long Coyote™; Dum Dum™; Elroy Jetson™; Fancy™; Fleegle™; Flintstones™; Fred Flintstone™; George Jetson™; Gravel Slag™; Goober™; Goober and the Ghost Chasers™; Hair Bear Bunch™; Hoppy the Hopparoo™; Huckleberry Hound™; Jetsons™; Josie and the Pussycats™; Judy Jetson™; Lippy the Lion™; Little Lambsey™; Magilla Gorilla™; Mildew Wolf™; Morocco Mole™; Motormouse™; Mr. Jinx™; Mumbley™; Mushmouse™; Muttley™; Officer Dibble™; Pa Rugg™; Pebbles™; Peter Potamus™; Pixie™; Punkin'™; Puss™; Quick Draw McGraw™; Ricochet Rabbit™; Rock Slag™; Rosie the Robot™; Scooby Doo™; Scrappy Doo™; Sebastian™; Secret Squirrel™; Shagg Rugg™; Sheepdog™; Snagglepuss™; Sneezly™; Snuffles™; Snorky™; So-So the Monkey™; Space Ghost™; Space Kidettes™; Space Pup™; Spook the Cat™; Square Bear™; Squiddly Diddly™; Top Cat™; Touche Turtle™; Wally Gator™; Wilma Flintstone™; Yipee™; Yogi Bear.
Harvey Comics, Inc.: BabyHuey™;Casper™;Fatso™; Herman™; Katnip™; Little Audrey™; Stinky™; Stretch™.
Hobo Joe Restaurants: Hobo Joe™.
John Stampone: Lil' Liberty Belle™.
Just-Rite Restaurants: Lil Miss JustRite™.
Kellogg's: Cornelius Sugarcoat™; Crackle™; Milton Toaster™; Pop™; Snap™; Sweetheart of Corn™; Tony the Tiger™.
King Features: KoKo the Clown™; Olive Oyl™; Popeye™; Swee' Pea™.
Kitty Pan Litter: Glamour Kitty™.
Larry Harmon: Bozo the Clown™; Laurel and Hardy™.
Maltex: Marky Maypo™.
Matt Groenig and Fox TV: Bart Simpson™; Homer Simpson™; Krusty the Clown™; Lisa Simpson™; Maggie Simpson™; Marge Simpson™; The Simpsons™.
Maurice Sendak: Creature™.
MGM/UA: Droopy™; Jerry™; Spike™; Tom™; Tom and Jerry™.
National Association of State Forests and the Forest Service, US Department of Agriculture: Smokey the Bear™; Woodsy Owl™.
Nestle: Hans™.
Nickelodeon: Ren and Stimpy™.
1980 Moscow Olympics/International Olympic Committee: Misha Olympia Bear™.
Paramount Pictures: Betty Boop™.
Pillsbury: Funny Face™.
Post Cereal: Linus Lionhearted™.
Quaker: Cap'n Crunch™; Jean Lafoote™; Quake™; Quisp™.
Quasar Industries: Quasar Robot™.
Renk Seed Company: Colonel Renk™.
Royal Gelatin: King Royal Gelatin Man™.
Sambo's Restaurants: Sambo™; Sambo's Tiger™.
Sterling Drug Company: Diaparene Baby™.
United Artists Pictures, Inc.: Pink Panther™.
United Feature Syndicate, Inc.: Charlie Brown™; Linus™; Lucy™; Pig Pen™; Sally™; Schroeder™; Snoopy™.
United Productions of America (UPA): Mr. Magoo™; Gerald Boing McBoing™.
Viacom International, Inc.: Attack of the Killer Tomatoes™; Deputy Dawg™; Heckle™; Jeckle™; Mighty Mouse™; Roquefort Mouse™.
Walt Kelly: Albert Aligator™; Beauregard Hound™; Churchy La Femme™; Howland Owl™; Pogo™; Porky Pine™.
Walter Lantz: Andy Panda™; Chilly Willy™; Oswald Rabbit™; Woody Woodpecker™.
Ward Productions: Boris™; Bullwinkle™; Dudley Ro-Right™; Fractured Fairy™; George of the Jungle™; Gidney Moonman™; Mr. Peabody™; Natasha™; Rocky™; Snidely™.
Warner Brothers: Animaniacs™; Bugs Bunny™; Bugsy™; Cool Cat™; Daffy Duck™; Dot™; Elmer Fudd™; Foghorn Leghorn™; Henery Hawk™; Honey Bunny™; K-9™; Looney Tunes™; Marvin the Martian™; Merlin the Magic Mouse™; Merry Melodies™; Muggsy™; Pepe Le Pew™; Pinky™; Porky Pig™; Road Runner™; Second Banana™; Sniffles™; Speedy Gonzales™; Sylvester™; Tasmanian Devil™; Taz™; The Brain™; Tweety™; Wakko™; Wile E. Coyote™; Yakko™; Yosemite Sam™.

Dedications

In loving memory of Grammy, who bought me more toys in my lifetime than anyone could fit into one book. She always wanted to see every new toy I got in the mail, and enjoyed them just as much as I did. And thanks to my wife, who kept telling me for years that I should "do a book."—Jameson Scott

This book is dedicated to my wife, Phyllis.—Jim Rash

Minnie Mouse™ small bank-6"-Dakin-1976-?-$6

Bambi™ vinyl figure-8"-Dakin-1974-BG-$30

CONTENTS

Introduction

Cartoon figural toys have been around almost as long as cartoons. The first cartoon was *Humorous Phase of Funny Faces* in 1906 by England's James Blackton. The first American animated film was Windsor McCay's *Gertie the Dinosaur* in 1909. Popular comic strips of the time, such as George Herriman's *Krazy Kat* and Bud Fisher's *Mutt and Jeff* were turned into cartoon shorts. Many of these newspaper favorites were captured in wooden, papier-mâché, and bisque figural toys, intended for their youthful readers.

Many famous animators started with silent cartoon shorts. Paul Terry, best known for Mighty Mouse, made *Aesop's Fables* starting in 1921. Max and Dave Fleischer, best known for Betty Boop and Popeye, started their *Inkwell Series,* starring Koko the Clown, in 1921. Walt Disney began in Kansas City with his *Alice* comedies, in 1924. Walter Lantz, best known for Woody Woodpecker, began with the dishonest *Colonel Heeza Liar*.

The silent era's most enduring character, Felix the Cat, was created by Pat Sullivan and Otto Messemer. This character launched a major wave of cartoon merchandise. Felix was a thoughtful, problem-solving feline who brought on an incredible wave of figural toys, which have stayed with us to this day.

After Walt Disney got burned with his character Oswald the Rabbit (he lost the rights to producer Charles Mintz), he went to great lengths to protect his next creation, Mickey Mouse. He carefully leased the rights of his now-famous mouse to many companies, and a merchandising bonanza was unleashed.

Other independently produced cartoon series emerged in the 1930's, including Leon Schlesinger's *Merry Melodies* and *Looney Tunes* (released by Warner Brothers) and the Fleischer brothers' brilliant rotoscoping (mixing live action backgrounds with hand-drawn animation) of Popeye and Betty Boop (released by Paramount).

The end of World War II signaled the release of many animators from the military. At this time, many studios seized control of their respective companies by producing each cartoon within their own studios. Warner Brothers tightened control of their *Looney Tunes/Merry Melodies* and worked in a backlot building nicknamed "Termite Terrace." Ultimately, Disney would return to making feature films.

The new medium of the post-World War II era, television, would herald new productions. Jay Ward created the first made-for-television cartoon in 1949: *Crusader Rabbit*, who battled evil-doers along with his cowardly sidekick, Rags the Tiger. Through the years, Jay Ward followed this TV series with *The Rocky and Bullwinkle Show, Hoppity Hooper,* and *George of the Jungle*. He would also create fully animated commercials for the advertising characters Cap'n Crunch, Quisp, and Quake.

Another influential animation studio was United Productions of America (UPA), creator of Mr. Magoo and Gerald McBoing Boing. This studio combined a minimalist background with simplified characters, to develop a surrealistic and cheaper process. This style would eventually spread to the rest of the industry.

When MGM closed their cartoon factory in 1957, William Hanna and Joseph Barbera, two of their animators, would be out of work. Having been the creative team behind *Tom & Jerry* for MGM, they moved on to form their own studio (Hanna-Barbera) and soon released their first animated television series, *Ruff and Reddy.* This was followed by *The Huckleberry Hound Show*, which featured a supporting character named Yogi Bear. Later characters developed by this most famous cartoon duo include Top Cat, The Flintstones, The Jetsons, and Scooby Doo.

One of the major companies to cover a wide range of characters and animation studios was R. Dakin and Company, San Francisco, California (hereafter, simply referred to as Dakin). We have also included advertising, Goofy Grams, and fictional character toys made by this company, due to their innovative design and sculpting work, and the fact that there is currently no other pictorial reference guide available for these wonderful figural toys. In the 1960's, Dakin made a "Tabasco the Bull" toy and other bean-filled plush toys, long before the current rage of bean-stuffed toys began in the mid-1990's.

Chapter 1.
How To Use This Book

This book concentrates on the post-World War II era, when space-age materials like polystyrene plastic allowed the mass production of many cartoon character figures. Included are toys from 1949 through the late 1990's. All characters are easily referenced by their respective animation company (Disney, Hanna-Barbera, Looney Tunes, etc.), which is then divided by character names. The only exception is under the Advertising Section's Dakin Goofy Grams and Dakin Banks, which can include Looney Tunes toys not shown under their own character's heading.

Figural toys in this book are mainly comprised of the following:

Plastic/vinyl jointed and non-jointed figures.
Plastic/rubber squeeze toys.
Plush dolls with vinyl heads and/or faces.
Plush and cloth dolls that are entirely stuffed.
Plastic figural shampoo bottles, called "soakies."
Plastic/vinyl figural banks.

Each character's set of photos will be followed by a detailed text identification listing, including the toy manufacturer's name, year of manufacture, height in inches (or length if the item is oblong), and the item's current value (based on the toy's scarcity, demand, and recent internet, toy shows, and toy publications sales). The value will be given for a loose, near-mint, complete toy. Of course, you must add to the value for toys in mint or pristine condition, and/or still in their original packaging, and subtract for toys that are incomplete or in a lesser grade condition. We will also list how each item was originally sold, using the following abbreviations: boxed (BX), bagged (BG), carded (CD), loose/sold as shown (L), or unknown (?).

Unless otherwise noted, captions will identify figures from left to right. Each item's identification line will contain information in the following order:

Description-Size-Manufacturer-Date-Packaging-Value

For example, the following caption looks like this:

Huckleberry Hound vinyl figure-7"-Dakin-1970-BG-$75

Please enjoy this collection of cartoon character figural toys. It is by no means complete, but we have tried to show a vast range of characters and figures. We have also tried to correctly identify and value everything as accurately as possible.

All information in this book is compiled from the authors' experience, knowledge, and belief.

We are not affiliated with any toy manufacturers or animation companies. We cannot guarantee that anyone will be able to buy or sell items shown in this book at the values quoted. These amounts are given only as a guide.

E-mail the authors (cartoonfig@aol.com) with any questions or comments.

Chapter 2.
Casper and Friends
(Harvey)

Baby Huey rubber figure-10"-Hungerford Plastics-1959-BG-$75
Katnip rubber figure-10"-Hungerford-1959-BG-$75
Herman rubber figure-8.5"-Hungerford-1959-BG-$75
Casper rubber figure-9"-Hungerford-1959-BG-$85

Casper rubber figure-7"-Hungerford-1960's-BG-$60
Casper rubber squeeze toy w/ puppy-8"-unknown-1960's-?-$50
Little Audrey rubber figure-13"-Hungerford-1959-BG-$75
Casper vinyl figure-7"-I.S. Sutton & Sons-1972-BG-$45

Casper vinyl puppet-6"-Pizza Hut premium-1995-BG-$4
Fatso vinyl puppet-7"-Pizza Hut premium-1995-BG-$4
Stinky vinyl puppet-7"-Pizza Hut premium-1995-BG-$4
Stretch vinyl puppet-7.5"-Pizza Hut premium-1995-BG-$4

Fatso vinyl figure-9"-Tyco-1995-BX-$5
Stinky vinyl figure-11"-Tyco-1995-BX-$5
Stretch vinyl figure-12"-Tyco-1995-BG-$5
Casper vinyl figure-9.5"-Tyco-1995-BG-$5

Katnip hand puppet-10.5"-Gund-1959-?-$50

Chapter 3. Disney

Donald Duck

Donald Duck plastic bank-5.5"-Dakin-1976-?-$8
Donald Duck plastic figure-6.5"-Disney park exclusive-1990-L-$6
Donald Duck bent leg vinyl doll-7.5"-Dakin-1978-BG-$15
Donald Duck straight leg vinyl doll-7.5"-Dakin-1974-BG-$25
Donald Duck plastic figure-5.5"-Dakin-1978-BG-$12

Donald Duck squeeze toy-6.5"-Dell-1960-BG-$30
Uncle Scrooge w/ nephews squeeze toy-7"-Dell-1960-BG-$60
Donald Duck stationary figure on stand-7.5"-Sutton-1972-BX-$20
Ludwig Von Drake twistables figure-6"-Marx-1962-BG-$30
Daisy Duck plastic bendie figure-5.5"-Applause-1991-L-$4

Huey vinyl figure w/ pink & blue hat-5"-Dakin-1976-BG-$25
Louie vinyl figure w/ pink & blue hat-5"-Dakin-1976-BG-$25
Dewey vinyl figure w/ pink & blue hat-5"-Dakin-1976-BG-$25
Louie vinyl figure w/ green & blue hat-5"-Dakin-1976-BG-$25
Huey vinyl figure w/ green & blue hat-5"-Dakin-1976-BG-$25
Dewey vinyl figure w/green & blue hat-5"-Dakin-1976-BG-$25

Huey orange bent leg figure olive shirt-5"-Dakin-1980-BG-$20
Huey orange bent leg figure green shirt-5"-Dakin-1980-BG-$20
Louie orange bent leg vinyl figure-5"-Dakin-1980-BG-$20
Dewey orange bent leg vinyl figure-5"-Dakin-1980-BG-$20
Huey yellow bent leg vinyl figure-5"-Dakin-1980-BG-$20
Louie yellow bent leg vinyl figure-5"-Dakin-1980-BG-$20
Dewey yellow bent leg vinyl figure-5"-Dakin-1980-BG-$20

Goofy

Goofy plastic figure-6"-Dakin-1978-BG-$10
Goofy plastic figure-7"-Dakin-1978-BG-$10
Goofy plastic figure-7"-Disney Park exclusive-1990-L-$6
Goofy vinyl doll (missing black vest)-9"-Dakin-1974-BG-$20

Lady and the Tramp

Tramp squeeze figure-4.5"-unknown-1960's-?-$10
Trusty vinyl figure-4"-made in Japan-1960's-?-$35
Lady vinyl figure-5"-made in Japan-1960's-?-$35
Lady latex squeeze toy-2.5"-Lanco (Spain)-1980's-BG-$15
Tramp vinyl figure-6.5"-made in Japan-1960's-?-$35
Tramp latex squeeze toy-4"-Lanco (Spain)-1980's-BG-$15

Mickey and Minnie Mouse

Mickey Mouse figure on stand-7"-Sutton-1972-BX-$20

Mickey Mouse plastic bank-6"-Dakin-1976-?-$8
Mickey Mouse plastic figure-5.5"-Dakin-1978-BG-$8
Mickey Mouse plastic figure-6.5"-Disney Park exclusive-1990-L-$6

Opposite page, bottom:
Minnie Mouse vinyl doll-7.5"-Dakin-1974-BG-$15
Mickey Mouse vinyl doll-7.5"-Dakin-1974-BG-$15
Mickey Mouse squeeze toy-6.5"-Holland Hall-1965-?-$15

Minnie Mouse small bank-6"-Dakin-1976-?-$6
Minnie Mouse plastic figure-5.5"-Dakin-1978-BG-$8
Minnie Mouse plastic figure-6.5"-Disney Park exclusive-1990-L-$6

Peter Pan

Tinkerbell figure on stand-8"-Sutton-1972-BX-$30
Captain Hook figure on stand-8"-Sutton-1972-BX-$30
Peter Pan figure on stand-8"-Sutton-1972-BX-$30

Peter Pan plastic figure-5.5"-Marx-1972-L-$10
Croc figural alarm clock-9.5"-Disney Store exclusive-1995-BX-$15
Tinkerbell plastic figure-5"-Marx-1972-L-$10

Pinocchio

Pinocchio squeeze toy-7.5"-Dell-1960-BG-$15

Pinocchio plastic figure-5.5"-Dakin-1978-BG-$15
Pinocchio vinyl doll removable shoes-8"-Dakin-1974-BG-$25
Pinocchio vinyl doll molded shoes-8"-Dakin-1974-BG-$25
Pinocchio doll-10"-Applause-1991-L-$10
Pinocchio plastic figure-6"-Disney Park exclusive-1990-L-$6

Pluto

Pluto latex squeeze toy-6"-Lanco (Spain)-1972-L-$15
Pluto rubber figure-5"-Combex (England)-1960's-?-$35

Pluto hard plastic figure-6.5"-Disney Park exclusive-1990-L-$6
Pluto vinyl figure-5"-Dakin-1978-BG-$12
Pluto orange vinyl figure-8"-Dakin-1974-BG-$20
Pluto yellow vinyl figure-8"-Dakin-1974-BG-$20

Snow White and the Seven Dwarfs

Dopey latex squeeze toy-5"-Lanco (Spain)-1985-L-$10
Grumpy latex squeeze toy-5.5"-Lanco (Spain)-1985-L-$10
Bashful latex squeeze toy-5"-Lanco (Spain)-1985-L-$12
Snow White vinyl bank-8"-Disney Park exclusive-1990-L-$6

Miscellaneous

Bambi vinyl figure-8"-Dakin-1974-BG-$30
Bongo (Fun & Fancy Free) jointed squeak toy-8"-unknown-60's-?-$40

Dumbo vinyl figure w/ moveable tail-7"-Dakin-1974-BG-$20
Dumbo vinyl figure w/ molded tail-5.5"-Dakin-1978-BG-$25

Winnie the Pooh vinyl figure-5"-Disney Park exclusive-1990-L-$6
Tigger vinyl figure-5.5"-Disney Park exclusive-1990-L-$6
Jungle Book Mowgli squeeze figure-6.5"-Holland Hall-1967-L-$25

Thomas O'Malley (Aristocats) plush-7"-sold with video-1995-BX-$8
Fantasia Magic Broomstick-8.5"-Disney store exclusive-1995-L-$6
King Louie (Jungle Book) sawdust filled doll-5.5"-Gund-1966-L-$30

Chapter 4.
Fictional, Advertising, and TV Characters

Cereal and Advertising Characters

Milton Toaster plastic bank-6"-Kellogg's mailaway-1974-L-$75
Cornelius Sugarcoat walker-6"-Kellogg's mailaway-1958-L-$200
Sweetheart of Corn plastic toy-8"-Kellogg's mailaway-1981-$15
Tony the Tiger squeeze toy-7"-Product People mailaway-1974-L-$30

Sonny Cocoa Puffs plastic bank-5"-General Mills-1991-L-$8
Lucky Charms plastic bank-4.5"-General Mills-1990-L-$8
Trix plastic bank-4.5"-General Mills-1990-L-$8
Cinnamon Toast Crunch plastic bank-5"-General Mills-1988-L-$8

Campbell Soup boy vinyl figure-7"-Product People-1974-BX-$20
Campbell Soup girl vinyl figure-7"-Product People-1974-BX-$20
Trix Rabbit vinyl figure-7"-Product People-1974-L-$35
Marky Maypo squeeze toy-10"-? (mailaway)-1960's-L-$40

Quisp figural bank-7"-Quaker mailaway-1972-L-$600
Cap'n Crunch vinyl bank-7.5"-Quaker mailaway-1974-$75
Speedy Alka Seltzer vinyl figure-8"-Bristol Meyer-1959-L-$500
Jean LaFoote vinyl bank-7.5"-Quaker mailaway-1974-$150

Linus Lionhearted plush doll-12"-Animal Fair mailaway-1964-L-$150
Hans-12"-Nestle Chocolate mailaway-1969-L-$40
Quisp pillow doll-11"-cereal mailaway-1968-L-$100
Quake pillow doll-10"-cereal mailaway-1968-L-$100

Funny Face cherry walker-3"-F & F Mold-1970-?-$75
Funny Face strawberry walker-3"-F & F Mold-1970-?-$75
Funny Face orange walker-3"-F & F Mold-1970-?-$75
Funny Face Goofy Grape walker-3"-F & F Mold-1970-?-$75

Frankenberry vinyl figure-8"-Product People-1974-BX-$200
Count Chocula vinyl figure-8"-Product People-1974-BX-$150
Fruit Brute vinyl figure-8"-Product People-1974-BX-$150
Boo Berry vinyl figure-8"-Product People-1974-BX-$160

Snap molded vinyl figure-8"-Kellogg's mailaway-1975-L-$20
Crackle molded vinyl figure-8"-Kellogg's mailaway-1975-L-$20
Pop molded vinyl figure-8"-Kellogg's mailaway-1975-L-$20

Snap moveable head vinyl figure-8"-Kellogg's mailaway-1975-L-$35
Crackle moveable head figure-8"-Kellogg's mailaway-1975-L-$35
Pop moveable head vinyl figure-8"-Kellogg's mailaway-1975-L-$35

Tony the Tiger bank-9"-Kellogg's mailaway-1967-L-$45
Twinkles plastic bank-8.5"-Transogram mailaway-1963-L-$400
Linus Lionhearted bank-10"-Transogram-1965-?-$300
King Royal Gelatin Man bank-9"-Niagra Plastics-60's-L-$150

Sambo's Tiger vinyl figure-6"-Dakin-1974-BG-$160
Sambo's Rest. Mascot vinyl doll-8"-Dakin-1974-BG-$125
Bob's Big Boy vinyl doll (missing burger)-8"-Dakin-1974-BG-$100
Lil Miss JustRite vinyl doll-7.5"-Dakin-1974('65 head reused)-$60
Lil Miss JustRite vinyl doll-7.5"-Dakin-1974('65 head reused)-$60

Misha Olympic Bear bank-5.5"-unknown-1980-?-$40
Quasar Robot Bank-7"-Dakin-1975-BG-$100
Smokey the Bear vinyl doll-8"-Dakin-1974-BG-$20
Bay View Eagle Bank-7"-Dakin-1976-?-$35

Glamour Kitty vinyl figure-7"-mailaway Dakin-1977-L-$175
Glamour Kitty figure(missing crown)-7"-mailaway Dakin-1977-L-$175
Glamour Kitty vinyl figure-7"-mailaway Dakin-1977-L-$175

Freddy Fast Gas figure (Douglas Oil Co.)-7"-Dakin-1976-BG-$50
Renk Seed Co. Colonel Renk vinyl doll-7"-Dakin-1970-BG-$300
Hobo Joe vinyl doll-8"-Dakin-1977-BG-$50
Lil' Liberty Belle vinyl doll-8"-Dakin-1975-BG-$50

Blue Bunny-7.5"-Dakin-19??-?-$75
Arabian vinyl figure-6"-Dakin-19??-?-$125
Woodsy Owl vinyl doll-7"-Dakin-1974-BG-$50
Diaparene Baby doll (Sterling Drug)-5"-mailaway Dakin-1980-L-$30

Smokey the Bear vinyl bank-6.5"-Dakin-1977-$35
Smokey the Bear small plastic figure-4.5"-Dakin-1974-CD-$20
Smokey the Bear vinyl figure-6"-Dakin-1976-BX-$35
Smokey the Bear squeakie toy-5"-Dakin-1974-L-$20

Dakin Banks (vinyl/plastic)

Speedy Gonzales on cheese-9"-1971-?-$25
Wile E. Coyote w/ bomb-12-1971-?-$35
Road Runner on hill-11"-1971-?-$25
Bugs Bunny on carrot basket-12"-1971-?-$25

Yosemite Sam on treasure chest 11"-1971-?-$35
Sylvester on fish crate-11"-1971-?-$20
Tweety on birdcage-9.5"-1971-?-$20
Tiger on cage-10.5"-1971-?-$15

Lion on cage-9.5"-1971-?-$15
Dog w/ flowers on doghouse-9.5"-1971-?-$15
Scotty on doghouse-10.5"-1971-?-$15
Seal-9"-1971-?-$20

Clown-6"-1971-?-$15
Humpty Dumpty-6"-1971-?-$15
"Davy Jones" the octopus-6"-1971-?-$15
"Crackerjack" mouse on cheese-6.5"-1971-?-$15

Monkey on barrel-10.5"-1971-?-$15
Bee on hive-7"-1971-?-$15
Smokey the Bear-10.5"-1971-?-$20

Mini HaHa vinyl doll-7"-1967-BG-$40
Tommy Hawk vinyl doll-7"-1967-BG-$35
Mercy Nursey vinyl doll blonde missing hat-7"-1968-BG-$50
Mercy Nursey vinyl doll black hair-7"-1968-BG-$50

Soccer Boy vinyl doll-8"-1970-BG-$20
Baseball Boy vinyl doll-8"-1970-BG-$20
Basketball (missing ball) vinyl doll-1970-BG-$20

Little Annette vinyl doll, dark brown hair-6.5"-1966-BG-$12
Little Annette vinyl doll, light brown hair-6.5"-1966-BG-$12
Little Annette vinyl doll, light brown hair-6.5"-1966-BG-$12
Little Annette vinyl doll, blonde hair-6.5"-1966-BG-$12
Little Annette vinyl doll, red hair-6.5"-1966-BG-$12

Little Annette vinyl doll, blonde hair-6.5"-1966-BG-$12
Little Annette white coat/red hair vinyl doll-6.5"-1966-BG-$12
Cowboy vinyl doll-6.5"-1965-BG-$25
Twinkie vinyl doll-7"-1967-BG-$35

Pigtail Anne vinyl doll, brown hair-7.5"-1965-BG-$20
Funny Valentine vinyl doll-6.5"-1965-BG-$75
Janie Christmas vinyl doll-7"-1966-BG-$75
Pigtail Anne vinyl doll, blonde hair-7.5"-1965-BG-$20

Angelica vinyl doll-7"-1966-BG-$25
Angel in sleepwear vinyl doll-7.5"-1967-BG-$25
Angelica vinyl doll (missing wings)-7"-1966-BG-$25

Dakin Goofy Grams (vinyl/plastic)

Speedy Gonzales-8.5"-1971-BX-$20
Wile E. Coyote w/ bomb-10.5"-1971-BX-$20
Merlin the Magic Mouse w/ martini glass-9"-1971-BX-$20
Dog w/ cloth clothing (tan)-9"-1971-BX-$20

Lion-7.5"-1971-BX-$12
Fox-8"-1971-BX-$12
Rufus Dog w/ heart (yellow)-6.5"-1971-BX-$25
Rufus Dog w/ heart (pink)-6.5"-1971-BX-$25

Kangaroo w/ baby (blue)-8"-1971-BX-$30
Kangaroo w/ baby (pink)-8"-1971-BX-$30
Graduation Dog-6.5"-1971-BX-$15
Bloodhound Dog-8.5"-1971-BX-$15

Dachshund Dog (brown; missing floral bouquet)-7.5"-1971-BX-$25
Christmas Mouse (missing songbook)-10"-1971-BX-$60
Road Runner-10"-1971-BX-$20
Dachshund w/flowers-7.5"-1971-BX-$20

Bull-7"-1971-BX-$10
Gorilla-7.5"-1971-BX-$10
Baseball Bear-8"-1971-BX-$25
Tweety w/ heart-8.5"-1971-BX-$25

Easter Bunny (missing flowers)-8.5"-1971-BX-$60
Elephant (purple)-8"-1971-BX-$25
Birthday Frog-8"-1971-BX-$40
Pepe Le Pew-9"-1971-BX-$60

Traveling Dog-9.5"-1971-BX-$30
Scottie Dog-7"-1971-BX-$20
Poodle-11"-1971-BX-$15
Bugs Bunny w/ hat and party favor-12"-1971-BX-$45

Mother Cat w/ kitten-8.5"-1971-BX-$25
Tiger-8"-1971-BX-$15
Tiger-8"-1971-BX-$15
Bowling Bear-8.5"-1971-BX-$25

Houston Astros bear(missing ball and banner)-9"-1971-BX-$85
San Diego Zoo Lion-8"-1971-BX-$85
Pepe Le Pew (missing railroad engineer's cap)-9"-1971-BX-$125

Dakin Squeakies

Bull squeeze toy-4"-1974-L-$8
Elephant squeeze toy-4"-1974-L-$8
Panda squeeze toy-4"-1974-L-$10
Polar Bear squeeze toy-5"-1974-L-$10
Monkey squeeze toy-4"-1974-L-$8
Bee squeeze toy-4"-1974-L-$8

Other Squeakies can be found in individual character groups.

Wolf Merry Menagerie vinyl figure-6.5"-Poynter-1968-L-$200
Horse Merry Menagerie vinyl figure-9"-Poynter-1968-L-$200

Dragon Merry Menagerie vinyl figure-9.5"-Poynter-1968-L-$200
Cow Merry Menagerie vinyl figure-6.5"-Poynter-1968-L-$200

Note: Revell model kit values are for boxed/unassembled kits-subtract 50% for assembled kits with box-subtract 70% for assembled kits with no box

Gowdy the Dowdy Grackle model figure-11"-Revell-1960-BX-$200
Norval the Bashful Blinket model figure-7"-Revell-1960-BX-$200
Tingo the Noodle-Topped Stroodle model kit-8"-Revell-1960-BX-$200

Busby Tasselated Afghan model kit figure-11"-Revell-1960-BX-$300
Grickily Gractus model kit figure-10"-Revell-1960-BX-$300
Roscoe the Many Footed Lion model figure-14"-Revell-1960-BX-$250

Cat in the Hat model kit figure-12"-Revell-1960-BX-$250

Cat in the Hat w/ Thing 1&2 model figures-12"-Revell-1960-BX-$400

Birthday Bird snap Model Kit-11"-Revell-1960-BX-$300

Horton the Elephant snap Model Kit-11"-Revell-1960-BX-$975

Hedwig plush doll-8"-Mattel-1970-BX-$250
Cat in the Hat talking plush doll-8"-Mattel-1970-BX-$100
Yertle the Turtle talking plush doll-8"-Mattel-1970-BX-$200

Horton the Elephant talking plush doll-
Mattel-1970-BX-$500

Creature plush by Maurice Sendak-13"-Determined-1980-L-$10
Creature plush by Maurice Sendak-13"-Determined-1980-L-$10

The Grinch plush doll-29"-Determined(Macy's Exclusive)-1997-L-$35

Laurel and Hardy

Oliver Hardy vinyl squeeze figure-6"-Dakin-1974-L-$20

Oliver Hardy vinyl doll-7.5"-Dakin-1974-BG-$35

Stan Laurel vinyl squeeze figure-6"-Dakin-1974-L-$20

Stan Laurel vinyl doll-8"-Dakin-1974-BG-$35

Pogo (Walt Kelly)

Howland Owl-4.5"-unknown (sold with detergent/soap)-1969-L-$10
Pogo-4"-unknown (sold with detergent/soap)-1969-L-$10
Porky Pine-unknown (sold with detergent/soap)-4.75"-1969-L-$10
Churchy La Femme-4.5"-unknown (sold with detergent)-1969-L-$10
Beauregard Hound-5"-unknown (sold with detergent/soap)-1969-L-$10
Albert Alligator-5"-unknown (sold with detergent/soap)-1969-L-$10

Attack of the Killer Tomatoes boy-7.5"-Applause-1991-L-$8

Attack of the Killer Tomatoes girl-7.5"-Applause-1991-L-$8

Attack of the Killer Tomatoes Mad Doctor-8"-Applause-1991-L-$8

Bozo the Clown squeakie-5.5"-Dakin-1974-L-$10

Colonel Capt. Kangaroo-10"-unknown (marked CBS TV)-1958-?-$50

Bozo vinyl figure-8"-Dakin-1974-BG-$18

Bump in the Night Molly figure-9.5"-Green Grass Prod.-L-$8
Bump in the Night Mr. Bumpity-8"-Green Grass Prod.-L-$8
Bump in the Night Squishington-7.5"-Green Grass Prod.-L-$8

Chapter 5.

Hanna-Barbera

Banana Splits

Bingo vinyl figure-7"-I.S. Sutton & Sons-1970-BG-$120
Snorky vinyl figure-7"-I.S. Sutton & Sons-1970-BG-$150
Drooper vinyl figure-7"-I.S. Sutton & Sons-1970-BG-$100
Fleegle vinyl figure-7"-I.S. Sutton & Sons-1970-BG-$100

Snorky flocked hair doll-7"-I.S. Sutton & Sons-1970-L-$125
Fleegle flocked hair doll-7"-I.S. Sutton & Sons-1970-L-$125
Bingo flocked hair doll-7"-I.S. Sutton & Sons-1970-L-$75
Drooper flocked hair doll-7"-I.S. Sutton & Sons-1970-L-$75

The Flintstones

Dino vinyl figure-10"-Knickerbocker-1960-BX-$70
Wilma Flintstone vinyl figure-12"-Knickerbocker-1960-BX-$95
Fred Flintstone vinyl figure-11"-Knickerbocker-1960-BX-$60

Baby Puss vinyl figure-10"-Knickerbocker-1960-BX-$125

Betty Rubble vinyl figure-12"-Knickerbocker-1960-BX-$95

Barney Rubble figure-10"-Knickerbocker-1960-BX-$60

Fred vinyl doll-8"-Dakin-1970-BG-$25
Fred King's Island plastic bank-6"-Dakin-L-$20
Barney vinyl doll-7"-Dakin-BG-$25
Barney King's Island vinyl figure-6"-Dakin-1980-L-$25

Bamm Bamm yellow hat (missing club and bone)-7"-Dakin-1970-BG-$75
Bamm Bamm orange hat-7"-Dakin-1970-BG-$35
Bamm Bamm latex squeeze toy-3.5"-Lanco (Spain)-1967-BG-$50
Bamm Bamm vinyl figure-5"-Applause-1991-L-$10

Pebbles vinyl figure-5"-Applause-1991-L-$10
Pebbles blonde hair Dakin-8"-Dakin-1970-BG-$35
Pebbles red hair original felt outfit-8"-Dakin-BG-$60
Pebbles red hair Dakin-8"-Dakin-1970-BG-$35

Dino vinyl figure-8"-Dakin-1970-BG-$50
Baby Puss vinyl figure-8"-Dakin-1971-$95
Hoppy the Hopparoo vinyl figure-8"-Dakin-1971-$95
Dino King's Island theme park vinyl figure-6"-Dakin-1982-L-$25

Dino light blue squeeze toy-6.5"-unknown-1962-?-$15
Dino King's Island squeeze figure-5"-Dakin-1982-L-$15
Dino with gold hanging dog tag-8"-Dakin-1970-BG-$65
Dino King's Island plastic bank-6"-Dakin-1982-L-$20

Fred vinyl figure-5.5"-Knickerbocker-1962-BX-$50
Fred latex squeeze toy-6"-Lanco (Spain)-1967-BG-$50
Barney latex squeeze toy-5.5"-Lanco (Spain)-1967-$50
Barney vinyl figure-5.5"-Knickerbocker-1962-BX-$50

Hoppy the Hopparoo push puppet-5"-Kohner-1967-BX-$75
Dino vinyl figure-5.5"-Knickerbocker-1962-BX-$50
Betty latex squeeze toy-5"-Lanco (Spain)-1968-BG-$40
Wilma latex squeeze toy-4.5"-Lanco (Spain)-1968-BG-$50
Fred King's Island vinyl figure-7"-Dakin-1980-L-$25
Fred Push Puppet-4.5"-Kohner-1960's-BX-$20

Pebbles Spencer Gifts movie doll-9"-Dakin-1994-BX-$10

Bamm Bamm Spencer Gifts movie doll-9"-Dakin-1994-BX-$10

Pebbles arcade plush doll-11"-Play-By-Play-1995-L-$10
Baby Puss plush doll-10"-unknown-1980's-L-$15
Bamm Bamm arcade plush doll-11"-Play-By-Play-1995-L-$10

Betty Rubble arcade plush doll-11"-Play-By-Play-1995-L-$10
Barney arcade plush doll-10.5"-Play-By-Play-1995-L-$10

Fred arcade plush doll-11"-Play-By-Play-1993-L-$10
Dino arcade plush doll-15"-Play-By-Play-1993-L-$10

Hair Bear Bunch

Bubi Bear flocked hair doll-6.5"-I.S. Sutton & Sons-1970-L-$50

Hair Bear flocked hair doll-8"-I.S. Sutton & Sons-1970-L-$60

Square Bear flocked hair doll-7"-I.S. Sutton & Sons-1970-L-$50

Bubi Bear latex squeeze toy-3.5"-Lambert Kay-1972-BG-$35
Square Bear latex squeeze toy-5"-Lambert Kay-1972-BG-$35
Hair Bear latex squeeze toy-5"-Lambert Kay-1972-BG-$35
Hair Bear plastic bank-7"-King's Four-1973-L-$75

Huckleberry Hound

Huck Hound King's Island vinyl figure-5.5"-Dakin-1982-L-$25
Huck Hound vinyl figure-7"-Dakin-1970-BG-$75
Huck Hound sawdust filled doll-6"-unknown-60's-?-$50
Huck Hound plastic bowling pin-7"-Transogram-early 60's-BX-$10

Huck Hound latex squeeze toy-5.5"-Lanco (Spain)-1967-BG-$50
Huck Hound latex squeeze toy-3"-Lanco (Spain)-1980's-BG-$10
Huck Hound King's Island squeeze toy-5"-Dakin-1982-L-$10
Huck Hound vinyl figure-5.5"-Knickerbocker-1962-BX-$40
Huck Hound squeeze toy-7"-Goebbels-1965-BG-$40
Huck Hound King's Island vinyl bank-5"-Dakin-1982-L-$12

Huck Hound plush doll-10"-Mattel-1993-BX-$8
Huck Hound arcade plush doll-10.5"-Plau-By-Play-1995-L-$10

The Jetsons

Rosie the Robot vinyl figure-6.5"-Bucky (Mexico)-1973-BG-$100

George Jetson vinyl doll-11.5"-Applause-1990-L-$10
Rosie the Robot vinyl doll-9"-Applause-1990-L-$10
Elroy Jetson vinyl doll-8"-Applause-1990-L-$10
Judy Jetson vinyl doll-9"-Applause-1990-L-$10

Rosie the Robot figure (white/gray top)-5.5"-Applause-1990-L-$8

Astro arcade plush doll-10.5"-Play-B-Play-1995-L-$10

Rosie the Robot figure (all white on top)-5.5"-Applause-1990-L-$8

Magilla Gorilla

Magilla Gorilla plush doll (vinyl head)-11"-Ideal-1964-BX-$125
Magilla Gorilla arcade plush doll-11"-Play-By-Play-1995-L-$10
Magilla Gorilla Target Barrel-10"-Ideal-1964-BX-$100

Magilla Gorilla plush doll (vinyl head)-7"-Ideal-1965-BX-$60
Magilla Gorilla latex squeeze toy-6"-Magvi (Spain)-1967-BG-$60
Magilla Gorilla vinyl figure-8.5"-?(Germany)-1982-BX-$75

Magilla Gorilla vinyl figure-8"-Bucky (Mexico)-1973-BG-$100
Magilla Gorilla latex squeeze toy-2"-Lanco (Spain)-19??-BG-$10
Magilla Gorilla push puppet-5"-Kohner-1969-BX-$25
Magilla Gorilla pull toy figure-7"-Ideal-1965-BX-$35

Magilla Gorilla plush doll-10"-Mattel-1993-BX-$8

Magilla Gorilla plush doll-17"-Presents-1985-L-$15

Mushmouse and Punkin' Puss

Punkin' Puss plush doll (vinyl head)-6"-Ideal-1965-BX-$50
Mushmouse plush doll (vinyl head)-6"-Ideal-1965-BX-$50
Punkin' Puss latex squeeze toy-5"-Magvi (Spain)-1967-BG-$40

Mushmouse latex squeeze toy-2"-Lanco (Spain)-19??-BG-$20
Mushmouse pull toy figure-6"-Ideal-1965-BX-$30
Punkin' Puss pull toy figure-6"-Ideal-1965-BX-$30
Punkin' Puss latex squeeze toy-2.5"-Lanco (Spain)-19??-BG-$20

Pixie and Dixie and Mr. Jinx

Pixie and Dixie squeeze figure-3.5"-Goebbels (Germany)-1965-BG-$40
Pixie and Dixie squeeze toy-4.5"-Dell-1960-BG-$85
Dixie incorrect female vinyl figure-5"-Knickerbocker-1962-BX-$50
Pixie vinyl figure-5"-Knickerbocker-1962-BX-$50

Mr. Jinx latex squeeze toy-3"-Lanco (Spain)-1980's-BG-$10

Mr. Jinx Push Puppet (missing label)-4"-Kohner-1960's-BX-$30

Mr. Jinx yellow figure-6"-Goebbels-1965-BG-$90

Mr. Jinx gray figure-8.5"-Bucky (Mexico)-1973-BG-$75

Quick Draw McGraw

Quick Draw McGraw vinyl figure-5.5"-Knickerbocker-1962-BX-$50

Quick Draw McGraw latex squeeze toy-8"-Lanco(Spain)-1980's-BG-$10

Quick Draw McGraw latex squeeze toy-9"-Dell-1960-BG-$100

Quick Draw McGraw sawdust doll -9"-unknown-60's-?-$45

Ricochet Rabbit and Droop-A-Long Coyote

Ricochet Rabbit plush doll (vinyl face)-12"-Ideal-1965-BX-$150

Ricochet Rabbit plush doll (vinyl head)-7"-Ideal-1965-BX-$65
Droop-A-Long Coyote plush doll (vinyl head)-7"-Ideal-1965-BX-$50
Droop-A-Long Coyote pull toy figure-7"-Ideal-1965-BX-$35
Ricochet Rabbit pull toy figure-7"-Ideal-1965-BX-$35

Droop-A-Long Coyote latex squeeze toy-6"-Magvi (Spain)-BG-$50
Ricochet Rabbit latex squeeze toy-5.5"-Magvi (Spain)-BG-$50
Ricochet Rabbit Push Puppet-4.5"-Kohner-1960's-BX-$30

Scooby Doo

Scooby Doo vinyl figure-8"-I.S. Sutton & Sons-1970-BG-$50
Scooby Doo flocked doll-7"-I.S. Sutton & Sons-1970-L-$35
Scooby Doo vinyl figure 7" W.B. store exclusive-1997-L-$8
Scooby Doo King's Island vinyl figure-8"-Dakin-1980-L-$75

Scooby Doo claw machine plush doll-8"-Play-By-Play-1990's-L-$12
Scooby Doo King's Island vinyl bank-5.75"-Dakin-1980-L-$75
Scrappy Doo King's Island vinyl figure-6"-Dakin-1982-L-$60
Scooby Doo Dream Pets cloth doll-8"-Dakin-1979-L-$30

Scooby Doo nodder-7.5"-WB store exclusive-1997-BX-$15

Scooby Doo all 4's King's Island PVC-4"-Dakin-1980-$15

Scooby Doo hard plastic Bank-7"-King's Four-1973-L-$75

Scooby Doo parachuter PVC missing chute-3"-Imperial-1977-CD-$15

Scooby Doo vinyl figure-7"-I.S. Sutton & Sons-1970-BG-$75
Scooby Doo squeeze figure-5"-Dakin-1980-$12
Scrappy Doo PVC figure-2.5"-Dakin-1982-$12

Benny the Ball Shoe Toss plastic figure-7"-Transogram-1962-BX-$50
Top Cat and Benny the Ball ramp walker-3"-Marx-1962-L-$85
Benny the Ball hand puppet-10"-Ideal-1963-?-$200
Fancy Shoe Toss Game plastic figure-7"-Transogram-1962-BX-$65
Top Cat Shoe Toss Game plastic figure-8"-Transogram-1962-BX-$40

Officer Dibble plush doll (vinyl head)-9"-Ideal-1962-BX-$75
Benny the Ball plush doll (vinyl head)-5.5"-Ideal-1962-BX-$200
Top Cat plush doll (vinyl head)-6.5"-Ideal-1962-BX-$125

Brain plush doll-14"-Presents-1985-L-$25
Top Cat arcade plush doll-10.5"-Play-By-Play-1995-L-$10
Top Cat plush doll-17"-Presents-1985-L-$20

Top Cat PVC figure-2.25"-Artoy (England)-1990's-L-$6
Benny the Ball PVC figure-1.5"-Artoy (England)-1990's-L-$6
Officer Dibble PVC figure-3"-Artoy (England)-1990's-L-$6

Wacky Races

Muttley PVC claw machine toy-3.5"-?(Japan)-1990's-L-$15
Dick Dastardly PVC claw machine toy-3.5"-?(Japan)-1990's-L-$15
Rock/Gravel Slag PVC claw machine toy-3.5"-?(Japan)-1990's-L-$15
Mumbly (Laff-A-Lympics) latex toy-4"-Lanco (Spain)-1975-BG-$40
Muttley plastic bank-7"-King's Four-1973-?-$60
Muttley plastic bank-7"-?(Japan)-1995-L-$20
Muttley latex squeeze toy-3.5"-Lanco (Spain)-1972-BG-$50

Muttley arcade plush doll-9.5"-Play-By-Play-1995-L-$10
Muttley claw machine plush doll-7"-Play-By-Play-1995-L-$12

Yogi Bear

Yogi Bear plastic bowling pin-7"-Transogram-early 60's-BX-$10
Yogi Bear squeeze toy-6"-Dell-early 60's-BG-$30
Yogi Bear vinyl figure-8"-Dakin-1970-BG-$60
Yogi Bear King's Island vinyl figure-7"-Dakin-1980-L-$15

Yogi Bear King's Island PVC figure-4"-Dakin-1980-L-$10
Yogi Bear King's Island bank-6"-Dakin-1980-L-$10
Yogi Bear cloth Dream Pets doll-7.5"- Dakin-1977-L-$20
Yogi Bear rubber squeeze toy-8.5"-Dell-1960-BG-$30
Yogi Bear King's Island squeeze toy-5"-1980-L-$15

Boo Boo latex squeeze toy-3"-Lanco (Spain)-1980's-BG-$10
Boo Boo latex squeeze toy-3"-Lanco (Spain)-1967-BG-$25
Yogi Bear latex squeeze toy-3"-Lanco (Spain)-1980's-BG-$10
Boo Boo squeeze toy-5"-Goebbels (Germany)-1965-BG-$40
Yogi Bear squeeze toy-6"-?(Germany)-1965-BX-$50
Yogi Bear latex squeeze toy-5"-Lanco (Spain)-1967-BG-$50

Boo Boo plastic figure-4.5"-store giveaway (Italy)-1967-L-$35
Papa Yogi Bear plastic figure-4.5"-store promo (Italy)-1967-L-$35
Yogi Bear plastic figure-4.5"-store promo (Italy)-1967-L-$35
Baby Yogi Bear plastic figure-4.5"-store promo (Italy)-1967-L-$35
Yogi Bear vinyl figure-5.5"-Knickerbocker-1962-BX-$50
Cindy Bear vinyl figure-5.5"-Knickerbocker-1962-BX-$50

Boo Boo claw machine plush doll-7.5"-Play-By-Play-1995-L-$12
Boo Boo arcade plush doll-9.5"-Play-By-Play-1995-L-$10
Yogi arcade plush Doll-10.5"-Play-By-Play-1995-L-$10
Yogi claw machine plush doll-7.5"-Play-By-Play-1995-L-$12

Miscellaneous

Atom Ant Tricky Trapeze-5"(fig. 2.5")-Kohner-1967-L-$55

Augie Doggie arcade plush doll-10.5"-Play-By-Play-1995-L-$12
Wally Gator arcade plush doll-10"-Play-By-Play-1995-L-$10

Wally Gator plush doll (vinyl head)-7"-Ideal-1962-BX-$150
Touche Turtle plush doll (vinyl head)-6.5"-Ideal-1962-BX-$125
Yipee Dog plush doll (vinyl head)-8"-Ideal-1965-BX-$55
Lippy the Lion plush doll (vinyl head)-7"-Ideal-1962-BX-$125

Peter Potamus plush doll (vinyl head)-7"-Ideal-1965-BX-$100
SoSo the Monkey plush doll (vinyl head)-6"-Ideal-1965-BX-$125
Sneezly Seal plush doll (vinyl head)-7"-Ideal-1965-BX-$85
Breezly Bruin plush doll (vinyl head)-7"-Ideal-1965-BX-$85

Dum Dum plastic soakie-8.5"-Purex-1960's-L (w/ paper art)-$50

Touche Turtle plastic soakie-9"-Purex-1960's-L (w/ paper art)-$40

Winsome Witch plastic soakie-10"-Purex-1966-L (w/ paper art)-$50

Snagglepuss plastic soakie-7.5"-Purex-1963-L (w/ paper art)-$50

Squiddly Diddly plastic soakie-10"-Purex-1966-L(w/ paper art)-$40

Secret Squirrel plastic soakie-10"-Purex-1966-L(w/ paper art)-$40

Morocco Mole plastic soakie-6"-Purex-1966-L(w/ paper art)-$90

Atom Ant plastic soakie-9"-Purex-1966-L(w/ paper art)-$75

Chopper plastic figure-4.5"-store giveaway (Italy)-1967-L-$35
Castor Beaver figure-4.5"-store giveaway (Italy)-1967-L-$35
Little Tom Tom figure-4.5"-store giveaway(Italy)-1967-L-$85
Fibber Fox plastic figure-4.5"-store giveaway (Italy)-1967-L-$35

Snooper sawdust filled doll-7"-unknown-60's-?-$65
Snagglepuss vinyl figure-7.5"-Dakin-1971-BG-$100
Hokey Wolf vinyl figure-8"-Dakin-1971-BG-$150

Secret Squirrel push puppet-4.5"-Kohner-1968-BX-$30
Precious Pupp rubber squeeze toy-5.5"-Magvi (Spain)-1967-BG-$150
Atom Ant push puppet-3.5"-Kohner-1969-BX-$30
Atom Ant latex squeeze toy-2"-Lanco (Spain)-1968-BG-$100
Goober (Goober & the Ghostchasers) bank-7"-King's Four-1973-L-$60

Pa Rugg latex squeeze toy-6.5"-Lanco (Spain)-1967-BG-$40
Breezly vinyl pull toy figure-6"-Ideal-1965-BX-$40
Wally Gator PVC figure-3"-Comico (Spain)-1991-?-$10
Peter Potamus PVC figure-3"-Comico (Spain)-1991-?-$10
Touche Turtle squeeze toy-5"-Bucky (Mexico)-1973-BG-$50

Sheepdog latex squeeze toy-5"-Lanco (Spain)-1972-BG-$30
Little Lambsey latex squeeze toy-3"-Lanco (Spain)-1972-BG-$30
Mildew Wolf latex squeeze toy-4.5"-Lanco (Spain)-1972-BG-$30
Augie Doggie plastic figure-4.5"-Bucky (Mexico)-1973-BG-$50
Snuffles vinyl figure-5.5"-Knickerbocker-1962-BX-$50
Snagglepuss plastic figure-4.5"-store giveaway (Italy)-1967-L-$35

Boo the Bull Dog latex squeeze toy-4"-Lanco (Spain)-1972-BG-$30
Spook the Cat latex squeeze toy-4"-Lanco (Spain)-1972-BG-$30
Chattanooga Cat latex squeeze toy-4"-Lanco (Spain)-1972-BG-$75
Blip (Space Ghost) latex squeeze toy-3"-Lanco(Spain)-1967-BG-$100
Shagg Rug latex squeeze toy-2.5"-Lanco (Spain)-1967-BG-$30
Autocat pilot latex squeeze toy-4"-Lanco (Spain)-1972-BG-$35
Motormouse latex squeeze toy-3.5"-Lanco (Spain)-1972-BG-$30
Space Pup (Space Kidettes)squeeze toy-3"-Lanco(Spain)-1973-BG-$30
Sebastian (Josie & the Pussycats)-2.5"-Lanco(Spain)-1972-BG-$35

Chapter 6.
Looney Tunes
(Warner Brothers)

Bugs Bunny

Bugs Bunny Bicentennial figure-9"-Dakin-1975-BG-$30
Bugs Bunny PVC figure-6"-Dakin-1976-L-$10
Honey Bunny PVC figure-4.5"-Dakin-1982-L-$12
Bugs Bunny vinyl figure-11"-Dakin-1971-BG-$20

ABOVE: Bugs Bunny hollow plastic figure-4"-Dakin-1974-CD-$10
Bugs Bunny beanbag body, vinyl head figure-10.5"-Dakin-?-$15
Bugs Bunny squeakie figure-6.5"-Dakin-1974-L-$20
Bugs Bunny vinyl bank-7"-Dakin-1976-?-$25
Bugs Bunny Cartoon Theater vinyl figure-7.5"-Dakin-1976-BX/BG-$20

BELOW: Bugs Bunny latex squeeze toy-7.5"Lanco (Spain)-1990's-BG-$10
Bugs Bunny latex squeeze toy-4"-Lanco (Spain)-1989-BG-$5
Bugs Bunny vinyl figure on stand-7.5"-Dakin-1971-?-$40
Bugs Bunny PVC figure-5.5"-Dakin-1986-L-$5
Bugs Bunny vinyl figure-11"-Six Flags exclusive-1990-L-$8
Bugs Bunny Marriott's Great America Bank-7"-Dakin-1975-?-$40

Bugs Bunny squeeze toy-7"-Oak Rubber-1949-BX-$35
Honey Bunny squeeze toy-6"-Oak Rubber-1949-BX-$75

Bugs Bunny Dream Pets cloth doll-8"-Dakin-1976-L-$10

Daffy Duck

Daffy Duck Cartoon Theater vinyl figure-7.5"-Dakin-1976-BX-$25

Daffy Duck Great America Bank-7"-Dakin-1976-?-$8

Daffy Duck vinyl figure-8"-Six Flags exclusive-1990-L-$8

Daffy Duck figure (missing white collar)-8.5"-Dakin-1968-BG-$25

Daffy Duck latex squeeze toy-6.5"-Lanco (Spain)-1988-BG-$10
Daffy Duck vinyl figure-9"-WB store exclusive-1995-L-$8
Daffy Duck vinyl figure-9"-Six Flags exclusive-1996-L-$8
Daffy Duck latex squeeze toy-4"-Lanco (Spain)-1988-BG-$5

Daffy Duck cloth Dream Pets doll-7.5"-Dakin-1976-L-$15

Elmer Fudd

Elmer Fudd in hunting outfit-8"-Dakin-1968-BG-$125

Elmer Fudd in tuxedo vinyl figure-9"-Dakin-1968-BG-$35

Elmer Fudd vinyl figure-7.5"-Dakin-1977-BG-$20

Elmer Fudd Marriott's Great America figure-4"-Dakin-1976-?-$20

Porky Pig

Porky Pig Squeeze toy-6.5"-Sun Rubber-1952-BG-$75
Porky Pig blue coat vinyl doll, cloth hat-7"-Dakin-1968-BG-$50
Porky Pig Squeeze toy-6"-Reliance-1978-BG-$10
Porky Pig black coat vinyl doll, plastic hat-7"-Dakin-1968-BG-$25

Porky Pig pink coat vinyl doll, cloth hat-7"-Dakin-1968-BG-$40
Porky Pig Squeakies squeeze toy-4"-Dakin-1974-L-$15
Porky Pig Birthday Gram-7"-Dakin-?-$45
Porky Pig black coat, cloth hat-7"-Dakin-1968-BG-$35
Porky Pig hollow plastic figure-3.5"-Dakin-1974-CD-$8

Porky Pig vinyl figure-8"-WB store exclusive-1995-L-$8
Porky Pig Cartoon Theater vinyl figure-6.5"-Dakin-1976-BX-$25
Porky Pig latex squeeze toy-3"-Lanco (Spain)-1990's-BG-$8
Porky Pig plastic bank-6"-Dakin-1976-?-$20
Porky Pig vinyl figure-7"-WB store exclusive-1993-L-$8

Road Runner

Road Runner vinyl/plastic figure-8.5"-Dakin-1968-BG-$25
Road Runner PVC figure-5"-Dakin-1976-?-$20
Road Runner plastic figure-4.5"-Dakin-1974-CD-$15
Road Runner Cartoon Theater vinyl figure-7.5"-Dakin-1976-BX-$35

Speedy Gonzales

Speedy Gonzales squeakies squeeze toy-5.5"-Dakin-1974-L-$20
Speedy Gonzales Cartoon Theater figure-7"-Dakin-1976-BX-$30
Speedy Gonzales latex squeeze toy-4"-Lanco (Spain)-1989-BG-$5
Speedy Gonzales message gram-7"-Dakin-1971-L-$30
Speedy Gonzales vinyl figure-7.5"-Dakin-1968-BG-$25

Sylvester

Sylvester vinyl figure-9"-WB store exclusive-1995-L-$8
Sylvester Cartoon Theater vinyl figure-7"-Dakin-1976-BX-$20
Sylvester latex squeeze toy-5.5"-Lanco (Spain)-1988-BG-$10
Sylvester vinyl figure-8"-Six Flags exclusive-1990-L-$8
Sylvester vinyl figure-7.5"-Six Flags exclusive-1996-L-$8
Sylvester plastic bank-7"-Dakin-1976-?-$20

Sylvester Stuffed cloth doll-8.5"-Dakin-1978-L-$10

Sylvester Plastic/vinyl Bank-11"-Dakin-1969-L-$25

Sylvester Squeeze toy-5"-Dakin-1974-L-$15

Sylvester PVC figure-4"-Dakin-1985-?-$8

Sylvester vinyl figure-8.5"-Dakin-1969-BG-$20

Sylvester rubber squeeze toy-6"-Oak Rubber-1949-BG-$50
Sylvester PVC figure-5"-Dakin-1976-?-$10
Sylvester PVC figure-5"-Dakin-1976-?-$10

Taz small PVC figure-4.5"-Dakin-1980-L-$10
Taz Great America bank-6"-Dakin-1976-?-$50
Taz vinyl figure-5"-Dakin-1978-BG-$400
Taz vinyl figure-7.5"-WB store exclusive-1995-L-$8
Taz small PVC figure-4"-Dakin-1985-?-$10

Taz vinyl figure-7"-Six Flags exclusive-1996-L-$8
Taz vinyl figure-6.5"-Six Flags exclusive-1990-L-$8
Taz talking action figure-6"-Tyco-1995-BX-$10
Taz latex squeeze toy-5"-Lanco (Spain)-1992-BG-$10

Tweety

Tweety Squeakies squeeze toy-5"-Dakin-1974-L-$5
Tweety vinyl figure-6"-Six Flags exclusive-1996-L-$8
Tweety vinyl figure-6"-WB store exclusive-1995-L-$8
Tweety message gram-6.5"-Dakin-1971-?-$30
Tweety vinyl figure w/ moveable arms-6.5"-Dakin-1968-BG-$15
Tweety PVC figure-3.5"-Dakin-1976-?-$10
Tweety PVC figure-3"-Dakin-1986-?-$10

Tweety vinyl figure-6.5"-Six Flags exclusive-1990-L-$8
Tweety hard plastic figure-3.5"-Dakin-1974-CD-$10
Tweety Cartoon Theater vinyl figure-6"-Dakin-1976-BX-$20
Tweety plastic bank-6"-Dakin-1976-?-$25
Tweety latex squeeze toy-4.5"-Lanco (Spain)-1980's-BG-$5

Tweety rubber squeeze toy-5"-Oak Rubber-1949-BX-$50

Tweety stuffed cloth doll-7.5"-Dakin-1976-L-$15

Wile E. Coyote

Wile E. vinyl figure-10"-Dakin-1968-BG-$15
Wile E. message gram vinyl figure-8.5"-Dakin-1971-?-$50
Wile E. Cartoon Theater vinyl figure-7.5"-Dakin-1976-BX-$25
Wile E. latex squeeze toy-7"-Lanco (Spain)-1988-L-$10
Wile E. PVC figure-6"-Dakin-1976-?-$10
Wile E. hard plastic figure-4.5"-Dakin-1974-CD-$8

Yosemite Sam

Yosemite Sam doll made w/o holsters & guns-7"-Dakin-1968-BG-$30
Yosemite Sam vinyl figure-6.5"-Dakin-1977-BG-$20
Yosemite Sam doll orange shirt/yellow scarf-7"-Dakin-1968-BG-$40
Yosemite Sam Squeakies squeeze toy-6"-Dakin-1974-L-$20

Yosemite Sam w/ black scarf missing hat/guns-7"-Dakin-1968-BG-$40

Yosemite Sam vinyl doll missing red scarf-7"-Dakin-1968-BG-$30

Miscellaneous

Cool Cat vinyl figure-9"-Dakin-1968-BG-$40
Foghorn Leghorn vinyl figure-8.5"-Dakin-1970-BG-$85
Pepe Le Pew vinyl figure-8"-Dakin-1971-BG-$85

Second Banana vinyl figure-6"-Dakin-1970-BG-$25

Merlin the Magic Mouse vinyl figure-7.5"-Dakin-1970-BG-$18

Marvin the Martian vinyl figure-9"-WB store exclusive-1995-L-$8
K-9 vinyl figure-8"-WB store excluseive-1995-L-$8
Marvin the Martian vinyl figure-9"-WB store exclusive-1992-L-$8
Marvin the Martian talking figure-6"-Tyco-1995-BX-$10

The Brain vinyl figure-6"-WB store exclusive-1995-L-$8

Wakko (Animaniacs) vinyl figure-8"-WB store exclusive-1996-L-$8
Dot (Animaniacs) vinyl figure-8"-WB store exclusive-1996-L-$8

Yakko (Animaniacs) vinyl figure-10"-WB store exclusive-1996-L-$8

Pinky (Animaniacs) vinyl figure-9.5"-WB store exclusive-1995-L-$8

Muggsy vinyl figure-7"-WB store exclusive-1995-L-$10

Bugsy vinyl figure-6"-WB store exclusive-1995-L-$10

Gossamer vinyl figure-6"-WB store exclusive-1994-L-$10

Gossamer vinyl figure-7"-WB store exclusive-1995-L-$10

Baby Bear vinyl figure-6.5"-WB store exclusive-1995-L-$10

Sniffles rubber squeeze toy-5.5"-Oak Rubber-1949-BX-L-$95

Henery Hawk latex squeeze toy-3.5"-Lanco (Spain)-1990's-BG-$10

Foghorn Leghorn PVC figure-4.5"-Dakin-1983-L-$10

Foghorn Leghorn latex squeak toy-5.5"-Lanco (Spain)-1990's-BG-$10

Chapter 7.
Other Cartoons

Alvin and the Chipmunks

Alvin squeeze toy-7"-Holland Hall-1964-?-$45
Alvin plastic soakie-10.5"-Colgate-1963-L (w/ paper art)-$15
Simon plastic soakie-10.5"-Colgate-1963-L (w/ paper art)-$15
Theodore plastic soakie-10.5"-Colgate-1963-L (w/ paper art)-$15

David Seville doll (sold with Chipmunks van)-8"-Ideal-1982-BX-$20

Beany and Cecil

Beany plush doll-7.5"-Caltoy-1984-L-$15
Cecil plush puppet-9"-Caltoy-1984-L-$30
Cecil soakie figure-7"-Colgate/Palmolive-1963-L(w/ paper art)-$45
Dishonest John pop-up figure-6.5"-Mattel-1962-BX-$15

Cecil talking plush doll-18"-Mattel-1961-BX-$60

Beany talking plush doll-15"-Mattel-1961-BX-$60

Mighty Mouse and Friends

Heckle/Jeckle rubber squeeze figure-7.5"-unknown-1958-BG-$75
Roquefort Mouse rubber squeeze figure-6.5"-unknown-1958-BG-$50
Mighty Mouse squeeze toy-5"-unknown-1980-5"-?-$15
Mighty Mouse plastic bank-5.5"-?(made in China)-1988-?-$5
Mighty Mouse rubber squeeze figure-9"-? (mailaway)-1958-$40

Deputy Dawg vinyl figure-6"-Dakin-1977-BG-$50
Mighty Mouse vinyl figure-6"-Dakin-1977-BG-$200

The Pink Panther

Pink Panther legs molded vinyl figure-8"-Dakin-1971-BG-$35
Pink Panther plastic figure-4"-Dakin-1974-CD-$15
Pink Panther Cartoon Theater vinyl figure-7.5"-Dakin-1976-BX-$25
Pink Panther vinyl figure-8"-Dakin-1971-BG-$30

Tom and Jerry and Friends (MGM)

Jerry rubber figure-1.5"-unknown-19??-?-$10
Droopy resin figure-6"-Demons and Marveilluse-1996-BX-$20
Droopy squeeze toy-4"-Alan Jay-1962-BG-$25
Tom rubber figure-2.5-unkown-19??-?-$10
Jerry latex squeeze toy-3"-Lanco (Spain)-1970's-BG-$5
Tom as Musketeer vinyl figure-6"-Bucky (Mexico)-1973-BG-$40
Spike vinyl squeeze toy-5.5"-unknown-1956-BG-$60
Tom PVC figure-3"-unkown-19??-?-$5
Jerry PVC figure-2"-unkown-19??-?-$5

Woody Woodpecker and Friends (Walter Lantz)

Chilly Willy rubber squeeze toy-6"-Oak Rubber-1951-BG-$60
Woody Woodpecker rubber squeeze toy-6"-Oak Rubber-1951-BG-$50
Andy Panda rubber squeeze toy-5.5"-Oak Rubber-1951-BG-$60
Woody Woodpecker vinyl bank-10"-IMCO-1977-BG-$30
Oswald Rabbit squeeze toy-7.5"-Sun Rubber-1952-BG-$45

Miscellaneous

Felix the Cat squeeze toy-6"-Irwin-1962-?-$50

Stimpy vinyl water squirt toy-5"-Dakin-1994-L-$10
Mr. Magoo vinyl figure-12"-unknown-1958-?-$150
Ren vinyl water squirt toy-5"-Dakin-1994-L-$10

Bart Simpson doll-7"-DanDee-1990-BG-$10
Maggie Simpson doll-4"-DanDee-1990-BG-$10
Homer Simpson doll-10"-DanDee-1990-BG-$10
Marge Simpson doll-13.5"-DanDee-1990-BG-$10
Lisa Simpson doll-7"-DanDee-1990-BG-$10
Krusty the Clown doll-10"-DanDee-1990-BG-$10

Chapter 8.
Peanuts Gang

Sally rubber squeeze figure-7.5"-Hungerford-1958-BG-$75
Charlie Brown rubber figure-9.5"-Hungerford-1958-BG-$35
Snoopy rubber squeeze figure-7.5"-Hungerford-1958-BG-$75
Linus squeeze figure(missing blanket)-8.5"-Hungerford-1958-BG-$85

Schroeder w/ piano rubber figure-7.5"-Hungerford-1958-BG-$250
Lucy rubber figure-9"-Hungerford-1958-BG-$40
Pig Pen rubber figure-9"-Hungerford-1958-BG-$100

Schroeder vinyl Pocket Doll-7"-Boucher-1966-BG-$50
Lucy vinyl Pocket Doll-7"-Boucher-1966-BG-$30
Linus vinyl Pocket Doll (missing blanket)-7"-Boucher-1966-BG-$40
Charlie Brown vinyl Pocket Doll-7"-Boucher-1966-BG-$30
Snoopy vinyl Pocket Doll-7"-Boucher-1966-BG-$40

Chapter 9.
Popeye and Friends
(King Features)

Olive Oyl Cartoon Theater vinyl figure-7.5"-Dakin-1976-BX-$30
Swee' Pea bean bag doll, vinyl head-7.5"-Dakin-1974-BG-$50
Popeye Cartoon Theater vinyl figure-7.5"-Dakin-1976-BX-$35

Olive Oyl vinyl doll-8"-Dakin-1974-BG-$35
Popeye vinyl doll, spinach can in right hand-8"-Dakin-1974-BG-$45
Popeye vinyl doll, spinach can in left hand-8"-Dakin-1974-BG-$25
Popeye Squeakies squeeze toy-5.5"-Dakin-1974-L-$20

Olive Oyl vinyl figure-6"-Applause-1990-L-$10
Swee' Pea vinyl figure-3"-Applause-1990-L-$10
Popeye plastic bank-5"-Dakin-1977-?-$35
Popeye vinyl figure-6"-Applause-1990-L-$10
Popeye hollow plastic figure-3.5"-Dakin-1974-CD-$15

Chapter 10.

Rocky and Bullwinkle and Friends

Gidney Moonman rubber figure-7"-Hungerford-1960-BG-$600
Sherman rubber figure-10"-Hungerford-1960-BG-$450
Mr. Peabody rubber figure-9.5"-Hungerford-1960-BG-$450
Bullwinkle rubber figure-12"-Hungerford-1960-BG-$900
Rocky rubber figure-8.5"-Hungerford-1960-BG-$550

Dudley Do-Right vinyl figure 7"-Dakin-1976-BX-$90
Bullwinkle vinyl figure-7"-Dakin-1976-BX-$75
Rocky vinyl figure-6.5"-Dakin-1976-BX-$75

Bullwinkle plastic soakie-11"-Colgate-1970-L (w/ paper art)-$20
Bullwinkle plastic soakie-11"-Colgate-1970-L (w/ paper art)-$30
Bullwinkle plastic soakie-11"-Colgate-1970-L (w/ paper art)-$40
Bullwinkle plastic soakie-11"-Colgate-1964-L (w/ paper art)-$50

Rocky push puppet-4"-Imperial-1972-CD-$150

Bullwinkle talking doll-10"-Mattel-1971-?-$100

Mr. Peabody push puppet-4"-Imperial-1972-CD-$150

Bullwinkle nodder-5"-Universal Studios-1993-L-$35
Bullwinkle bank-10"-IMCO-1977-?-$40
Rocky soakie-9"-Colgate-1964-L (w/ paper art attached)-$20
Fractured Fairy nodder-5"-Universal Studios-1993-L-$35

Sherman bendie-4"-Wham-O-1972-CD-$15
Mr. Peabody bendie-4"-Wham-O-1972-CD-$12
Natasha bendie-5.5"-Wham-O-1972-CD-$10
Boris bendie-3.5"-Wham-O-1972-CD-$10
Snidely bendie-5"-Wham-O-1972-CD-$15

Bullwinkle bendie-5"-Wham-O-1972-CD-$10
Rocky bendie-4"-Wham-O-1972-CD-$10
Dudley Do-Right bendie-5"-Wham-O-1972-CD-$12
Dudley Do-Right's horse bendie-5"-Wham-O-1972-CD-$100

CHAPTER 11.
UNDERDOG AND FRIENDS

Underdog bendie figure-5"-Jesco-1985-CD-$25
Underdog vinyl bank-7"-Street Kids-1995-BX-$50
Underdog Cartoon Theater-6"-Dakin-1976-BX-$125

Underdog vinyl bank-9.5"-IMCO-1977-BG-$75
Underdog vinyl bank-10"-Playpal-1973-BG-$60
Underdog vinyl bank-7"-Playpal-1973-BG-$50

King Leonardo plush doll-10.5"-Holiday Fair-1962-?-$300

Tooter the Turtle plush doll-10"-Holiday Fair-1962-?-$300